DK ESSENTIAL SCIENCE

P9-DJL-874

artificial intelligence

JACK CHALLONER

SERIES EDITOR JOHN GRIBBIN

LONDON, NEW YORK, MUNICH,
MELBOURNE, and DELHI

series editors Peter Frances, Hazel Richardson
DTP designer Rajen Shah
picture librarians Gemma Woodward,
Romaine Werblow
illustrator Richard Tibbitts

category publisher Jonathan Metcalf
managing art editor Philip Ormerod

Produced for Dorling Kindersley by
Design Revolution, Queens Park Villa,
30 West Drive, Brighton, East Sussex, BN2 0QW

editor Liz Wyse
designer Simon Avery

First American Edition, 2002

02 03 04 05 10 9 8 7 6 5 4 3 2 1

Published in the United States by
DK Publishing, Inc.
375 Hudson Street
New York, NY 10014

Library of Congress Cataloging-in-Publication Data

Challoner, Jack.
   Artificial intelligence / Jack Challoner.
     p. cm. -- (Essential science)
   Includes bibliographical references and index.
   ISBN 0-7894-8920-1 (alk. paper)
   1. Robotics. 2. Artificial intelligence. I. Title. II. Series.

TJ211 .C5124 2002
006.3--dc21
                                                     2002073562

Color reproduction by Colourscan, Singapore
Printed and bound by Graphicom, Italy

See our complete product line at www.dk.com

contents

what is artificial intelligence?

Modern computers come in a variety of guises, and can carry out a range of tasks. Many people own computers that can take dictation or automatically check a typed document for spelling mistakes; chess-playing computers can beat the world's Grand Masters; autonomous computer-controlled robots can explore other planets with minimal input from space flight engineers on Earth. Before the age of digital computers, all of these tasks could only have been carried out by people—does this mean that these devices are intelligent? In the future, might computers actually be conscious of what they are doing? Might there one day exist true electronic brains, with free will, emotion, and even a sense of morality? What use might such computers have, and what threats might they present? These questions are considered in a branch of science called "artificial intelligence."

independent travel

Among the most impressive examples of artificial intelligence are computerized roving vehicles that are designed to explore other planets, making decisions based on their surroundings and goals.

smart machines?

How was the party last night?

If you own a personal computer, you may have found yourself addressing it as if it were aware of your needs—praising it when it pleases you and shouting abuse at it when it does not. PCs and other electronic devices are unique. They are designed to be user-friendly and to do what you want.

In fact, one of the main drives toward creating artificial intelligence is the desire for more user-friendly technology. Other reasons for creating artificial intelligence include the need for more complete automation—so that intelligent robots can completely relieve human beings of repetitive or dangerous tasks; undertake the intelligent analysis of huge amounts of information; and enhance entertainment—so that computer games might seem more real, for example.

human machines
It is not uncommon to hear people converse with inanimate objects, even speaking about them in human terms.

electronic brains

Today's personal computers respond to information they receive—such as text input via a keyboard—according to sets of instructions called programs. By following programs, computers can behave as if they are intelligent. If powerful computers are used to control other devices, the illusion of intelligence can be even more convincing. There are computers that can drive cars and even fly aircraft, while a variety of robots carry out complex tasks in many different settings.

without windows
The US Air Force "ALTUS," once given its orders, can move out of its hangar, carry out surveillance flights transmitting video and other information, and return safely to its hangar—all with no human control.

ALTUS

The first digital electronic computers—built during the 1940s—were programmed laboriously, by changing the wiring of their circuits. But within a decade, computers could store and run programs, and process information automatically. The invention of the automatic computer encouraged many people to consider that machine intelligence was actually a possibility, and the science of artificial intelligence was given its name in 1956.

> If the human brain were so simple that we could understand it, we would be so simple that we couldn't. **"**
>
> ...erson Pugh (1977)

Modern computers are intelligently programmed, by intelligent people, to behave intelligently. Does that make them in some way "intelligent"? There is no clear definition of intelligence, but there are many behaviors or abilities that seem to require intelligence. Reasoning, prediction, empathy, and the ability to deal with new situations are important examples. Our own intelligence is a product of our brains. Some people think of the brain as an incredibly complicated computer: carrying out information processing on a grand scale; running a "program" that we call the mind; and responding to input it receives from the senses. Perhaps if computers were more powerful, and their programs more complex, they would become "electronic brains" with genuine intelligence.

The term "artificial intelligence" was coined in 1956 by American computer scientist **John McCarthy** (born 1927). A study at Dartmouth College set out to examine the idea that "every aspect of learning or any other feature of intelligence can in principle be so precisely described that a machine can be designed to simulate it." Now a senior fellow at Stanford University, California, he has written many programs that attempt to simulate intelligence.

big thinker?
In 1949, food processing company J. Lyons introduced a computer with stored programs into service. The EDSAC, as it was called, was the very first business computer. The name is an acronym for "Electronic Delay Storage Automatic Calculator."

demonstrating intelligence

Suppose that next week a computer programmer claimed that he or she had written a program that gives computers real intelligence. How would we evaluate this claim?

In 1950, British mathematician Alan Turing devised a scenario involving a human "interrogator" posing questions to both a computer and a human being, via a computer keyboard. The interrogator would receive replies via a computer screen or printer, and would figure out who was answering. Turing argued that a computer that could consistently fool the interrogator into believing it was the human being would be intelligent. He predicted, wrongly, that by 2000 computers would pass this test. Turing's scenario forms the basis of an annual competition, which aims to inspire researchers to expand the boundaries of artificial intelligence.

thinking prize
The Loebner Prize will be awarded to a computer program that gives an examiner consistently intelligent replies.

Another annual competition for intelligent machines is the RoboCup, in which teams of robots play football against each other, with no human intervention. The ultimate goal of the RoboCup organization is to have developed by the year 2050 "a team of fully autonomous humanoid robots that can win against the human soccer champions."

dogged competitor
Some of the RoboCup competitors make use of an existing robot— an electronic dog called AIBO—as a team player.

A computer passing the Turing Test or robots achieving the goal of winning the RoboCup would undoubtedly be incredible achievements, but not everyone is convinced that even these machines would necessarily be "thinking."

the turing test

The Turing Test normally involves judges posing questions to several computer programs and at least one human being. Many people have written computer programs that can analyze questions and create well-constructed replies—but to date none has come close to fooling the judges. The questions posed by the judges may be on any topic, and are generally intended to stimulate two-way conversation. They often include topics such as art, current affairs and, of course, artificial intelligence. In 1990, American philanthropist Hugh Loebner promised a prize of $100,000 and a solid gold medal to the designer of the first computer to pass the Turing Test. Loebner Prize competitions have been held every year since 1991. There is little prospect of the prize being claimed in the near future.

> Can you write me a short poem about an elephant?

an examiner poses questions and attempts to distinguish between the computer and the person

the computer program responds to the questions

a person at a computer terminal responds to the same questions

a true test?
Alan Turing believed that a computer program that gave consistently intelligent replies would necessarily be intelligent. Was he right?

British mathematician **Alan Turing** (1912–1954) is acclaimed as one of the most brilliant and influential thinkers of the 20th century. In the 1930s, he investigated the logical foundations of mathematics and physics, and played a vital role in the development of the electronic computer. He is best known for his work on code-breaking during World War II.

understanding words

In his article *Minds, Brains and Programs*, written in 1980, American philosopher John Searle argued convincingly that computers can never be truly intelligent because they will never be able to understand anything. He based his argument on the operation of computer programs that can answer comprehension questions about simple short stories. These programs, pioneered in the 1970s, are an example of "natural language processing."

short stories
A restaurant "script," designed to enable a computer to break down a story into basic concepts, will include actions such as ordering food, eating it, and paying the bill—or not.

They work by analyzing sentences, using the rules of grammar, and break down a simple story into its elements using "scripts"—preprogrammed outlines that apply to certain situations. Consider this short story: "Jack went to a restaurant and ordered an omelette with french fries. The waiter told him there were no french fries. Jack decided to have omelette with salad instead. He ate his food, but left without paying the bill."

A suitable computer program, which has been fed this story along with a restaurant script, can produce correct answers to questions such as: "Did Jack eat french fries?"; "Why did Jack eat salad instead of french fries?"; and "Was Jack satisfied with his meal?" Although these programs are very limited in their abilities, natural language processing was considered by some as a first step toward true artificial intelligence. Some researchers argued that because the replies given by these programs are indistinguishable from the answers a human respondent might give, the computer can actually be said to understand, just as people do.

what does it all mean?

Searle's argument against these claims centers on the difference between syntax (grammar) and semantics (meaning). The computer programs manipulate electronic pulses that represent the syntax in the short stories, but totally lack semantics. Human minds, on the other hand, understand semantics because they have a "feeling" about the things they perceive. Philosophers call this "intentionality." Intentional states relate to elements of the world around them, and attach meaning to them. They include beliefs, desires, fears, and pain. Without intentionality, a computer will not actually understand a story presented to it, despite the fact that it can answer questions about it correctly. As Searle points out, any intentionality that a computer program appears to have is found within the mind of the person who wrote the program.

story time
Even young children attach meanings to events in stories, which help them to understand what is going on. Computers, on the other hand, do not actually understand anything.

the chinese room

To illustrate his argument, John Searle put himself in the role of a computer carrying out natural language processing. In what has famously become known as the Chinese Room Argument, Searle placed himself in an imaginary locked room, surrounded by cards upon which are printed Chinese symbols. Through a slot in the door of the room, he is fed a story in Chinese, and then a series of cards that constitute questions about the story—again, in Chinese. He finds the correct symbols to respond to the questions, by following complex sets of instructions, and he passes sets of cards out of

the room. It appears to people outside the room that he understands Chinese. However, as Searle readily admits, he understands no Chinese at all.

no understanding
A computer program designed to answer questions simply follows instructions, giving the illusion of understanding.

Searle's argument seriously undermines the validity of the Turing Test, and strikes at the heart of the deeply held belief of many contemporary researchers in artificial intelligence—that computers may one day be truly intelligent. Since the publication of Searle's article, supporters of the claims of artificial intelligence have constructed many replies in defense of their beliefs. For example, the so-called Systems Reply insists that, taken as a whole, the system—consisting of Searle and the complicated sets of instructions which he follows—can be said to understand Chinese. Another argument, which has been dubbed

A digital computer can simulate everything—and simulated thinking the same as thinking!"

Marvin Minsky (interview for Generation5)

the Connectionist Reply, suggests a modification of the original analogy that Searle put forward. Instead of one person in the room manipulating Chinese symbols, imagine many people, each one processing only some of the commands. In this case, none of the individuals in the room understands what he or she is doing, but collectively, they can be said to understand Chinese. The debate is still raging.

group intelligence
A large colony of leaf-cutter ants displays collective intelligence. Might this phenomenon provide an answer to Searle's challenge?

what's in a word?

Was Alan Turing right to say that if a computer behaves intelligently, it can accurately be called intelligent? Or is John Searle right to insist that real intelligence requires an internal world of thoughts and feelings—a mind—and that this is something which is impossible to simulate in a computer? It is helpful to consider what the phrase artificial intelligence might actually mean.

Compare the meanings of the word "artificial" in the phrases "artificial turf" and "artificial flavorings." Artificial turf is not turf at all, but looks similar to real turf and shares some of its characteristics. If the "artificial" in artificial intelligence has this meaning, then a computer only has to share some of the

the real thing?
In what sense is artificial intelligence artificial? Is it the real thing, made artificially—like an artificial flavoring— or does it merely pretend to be the real thing—like artificial turf?

key points

• Computers can be programmed to behave intelligently

• There are strong philosophical arguments over the possibility of genuine artificial intelligence

• A distinction can be made between "Weak AI" and "Strong AI."

characteristics of human intelligence, but does not actually have to be intelligent. John Searle defines this as "Weak AI," and believes that even a computer that can pass the Turing Test would still only possess Weak AI.

Artificial flavorings in processed foods, on the other hand, really are flavorings, but are produced artificially in chemists' laboratories, rather than taken from nature. If the "artificial" in artificial intelligence has this meaning, then a computer must possess real intelligence, just produced in a different way from our own.

Such a computer would be able to think for itself—it would have a mind. Searle called this idea "Strong AI," and claims that it is impossible to achieve.

So, the ambiguity that the word "artificial" possesses usefully divides the artificial intelligence community into two clear camps: those who want to design useful computer systems and those who want to create artificial minds. When it comes to the word "intelligence," however, ambiguity is a major problem.

weak AI
A computer can be programmed to play chess without needing a mind of its own; this is an example of Weak AI.

strong AI
Examples of strong AI are found only in science fiction. Robby the Robot, from the film Forbidden Planet *(1957), has a high IQ and impeccable morals.*

thinking about thinking

Philosophers and scientists have long puzzled over the nature of intelligence. People who are described as intelligent may be skillful, knowledgeable, alert, witty, or sympathetic. Computers can carry out tasks that require skill, they can store information and are considered alert if they respond to certain stimuli. But at present, they can only do these things if they are programmed to do so. Furthermore, it would be very hard to imagine a computer that can be sympathetic or witty. Any definition of intelligence must address most or all of the following human abilities: reasoning, learning, judgment, remembering, emotion, intention, understanding, common sense, and consciousness. Computers can be programmed to learn and remember, but computers with judgment, emotion, intention, understanding, and common sense seem much more of a challenge. The greatest mystery of all is consciousness—the feeling of being aware of yourself and the world around you.

emotional robot
Kismet, an expressive robot designed at the Massachusetts Institute of Technology, is part of an attempt to make robots interact naturally with humans.

One thing that is certain is that—in humans at least—all of the qualities and abilities listed above seem to be possible because we have minds. Perhaps for a computer to be truly intelligent, it too would need a mind. The roots of philosophical ideas about the mind lie in Ancient Greece. Most of the Greek thinkers believed the mind to be located in the heart. Only a few realized the importance of the brain; the ancient Greek philosopher Plato was among them. However, Plato believed that the mind was completely separate from the physical world—an idea called dualism—and not caused by processes in the brain. The 16th-century French mathematician and philosopher, René Descartes, is another famous dualist. In one experiment, he showed how the lens inside a sheep's eye formed an image on the retina. He traced the optic nerve deep into the brain, where he believed that the "mind's eye" viewed the image. Many philosophers still adhere to dualism to this day. If it is true, it may be an insurmountable challenge—it is hard to imagine a soul being programmed into a computer.

In the late 19th century, the new scientific discipline of psychology aimed to develop scientific theories about the mind, to explain how we perceive the world and how we learn and remember, and to understand human and animal behavior. In the early 1900s, Austrian psychologist, Sigmund Freud formulated theories about the inner world of unconscious drives and dreams. Intelligence quotient (IQ) tests, introduced around the same time, attempt to measure what psychologists refer to as general intelligence. They gauge "cognitive" abilities, such as verbal reasoning and spatial and arithmetic skills. But exactly how the brain created

the mind's eye
Descartes showed how the lens of the sheep's eye formed an image on the retina, and proposed that the mind then created an image of its own.

"true" image re-created in the sheep's visual cortex

the inverted image forms on the retina

cognitive abilities or unconscious desires was a mystery. In the 1930s, neuroscience—the study of nerve cells, or neurons—began to make some headway. One of the most important goals of neuroscience is to understand how neurons in the brain contribute to aspects of intelligence such as perception, learning, and memory. Applying that understanding to computer system design is a promising approach to artificial intelligence (see p. 37).

understanding neurons

Neurons are the basic units of the nervous system. Each one consists of a small cell body with long, fibrous extensions called axons and dendrites (see pp. 18–19). In general, the ends of the axons of one neuron meet the ends of dendrites of many others. Inside the brain they form an incredibly complex and ever-changing network. The neurons whose fibers meet constantly exchange electrical signals. It has been estimated that there are as many as 100,000 million neurons in the brain—more than ten times the number of people on Earth. It has also been calculated that there are about 1,000,000,000,000,000 (one quadrillion) links between them. There is compelling evidence that our capacity to learn and remember is a result of the way neurons connect together, and it is almost certain that the signals produced by neurons form the basis of all our perceptions, thoughts, and behaviors. Most neuroscientists believe that even our consciousness emerges directly from these countless tiny nerve signals.

Neurons can be in one of two states: "firing" or "not firing." When a neuron is firing, it produces between about 50 and 100 pulses of electricity every second; when it is not firing, it produces only a few. A neuron fires if it receives

firing on all neurons
With millions of neurons firing in the brain at any one time, keeping them all under control might be an impossible juggling act. But the brain has evolved to deal with enormously complex tasks—including juggling.

enough stimulation from all the neurons to which it connects. Some neurons—those whose dendrites form our sense organs—receive their stimulation from the outside world. The skin, for example, contains the dendrites of sensory neurons that are sensitive to pressure and to heat, while dendrites in the eye are sensitive to light. The outputs of the human brain are carried by "motor

the role of neurons

Each neuron is surrounded by a membrane whose surface carries a voltage, or "potential." Changes in this voltage form the basis of nerve signals. A neuron produces electrical pulses that travel from the cell body along the axon. At the ends of the axon's many branches, the pulses are picked up by the dendrites of other neurons across tiny gaps called synapses.

dendrite

nucleus

electrical signal

cell body

myelin sheath

membrane

axon

Each pulse that reaches a synapse releases chemicals, called neuro-transmitters, which pass across the synapse and through the membrane of the dendrite. In turn, this affects the potential of the receiving neuron's membrane.

neurons," away from the brain to muscles. An electric pulse reaching a muscle will cause that muscle to contract. So, all our perceptions of the world—from hearing thunder to feeling the pain of a pinprick—begin as electric pulses at the ends of our sensory neurons. And the electric pulses in our motor neurons are responsible for everything from winking an eye to driving a car.

The receiving neuron has many dendrites, each receiving signals across a synapse; the neuron will fire if the overall change in its potential exceeds a certain threshold. As we experience and react to the world around us, new neurons and new synapses form, some old ones are lost, and the nature of a particular synapse can change.

axon terminal

synaptic junction

the synapse

The gap between the axon of one neuron and a dendrite of another is tiny—typically about 78 one-hundred thousandths of an inch (two ten-thousandths of a millimeter). Each type of neurotransmitter molecule has dedicated "receptor sites" on the membrane of the receiving dendrite.

an organized brain

The wrinkled, outermost part of the brain is called the cortex. There is an area of cortex on each side of the brain at which sensory signals arrive, and another from which motor signals originate. Signals from a particular part of the body, passing along sensory neurons, always end up in the same part of the "sensory cortex." Likewise, signals originating from a particular region of the "motor cortex," passing along motor neurons, always arrive at the same part of the body. This localization of function is the norm within the brain. It is as if the brain is precisely hardwired according to a complex master plan. The brain's architecture does follow a design that is common to all of us, but at the level of individual neurons and groups of neurons, we are all different. Our experiences mold our brains, by changing the connections between its neurons.

a place for everything
Specific areas of the brain have specific functions (as labeled right). Many of our higher functions, such as language and perception, seem to be located in the outer part of the brain: the cortex.

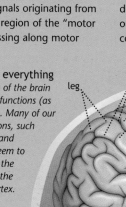

leg · torso · arm · hand · face · frontal cortex

body senses

visual

auditory

gustatory

olfactory

motor

computer brain?

For many neuroscientists, the brain is nothing more than a very complicated computer. And, in fact, there are many similarities between brains and computers. For example, both receive information in the form of electric pulses, process that information internally, and produce electric pulses as output. And even neurons have their counterparts inside digital computers: electronic circuits called logic gates. Just as neurons can be either firing or not firing, logic gates can be either "on" or "off." This two-state, or binary, system is at the heart of the way computers process information. The similarities do not stop there. Logic gates produce output (are "on") only when certain criteria are met in the patterns of their inputs—this is also true for neurons in the brain. Furthermore, most logic gates receive their inputs from other logic gates, forming networks as neurons do inside the brain. As with neurons, the exceptions are logic gates that either receive input from the world outside the computer or those that produce output.

Despite the similarities between brains and computers, and between neurons and logic gates in particular, there are important differences. Perhaps the most obvious is complexity: not even the most powerful computers have anywhere near as many logic gates as the brain has neurons. Another important difference is that the brain does not have to be programmed to do what it does. Perhaps most importantly, computer scientists understand exactly how computers work, while there is still much to discover about the brain.

hard wired?
The similarities between the brain and a computer are clear, but where does the analogy break down?

the inner world

recap

Our perception of the world arises as a result of the firing of neurons in the brain. Our capacity to learn is a result of the changing connections between neurons. An individual neuron "decides" to fire based on stimulation it receives from the firing of many other neurons.

Modern neuroscience has matched particular structures and regions of the brain with their functions, explained very clearly how individual neurons work, and has even helped us to understand how the connections between neurons can form memories and produce learning. But no one really understands the features of human intelligence that are very probably the most important to research into artificial intelligence— the production of meaningful language, intentional states and consciousness, and self-awareness. Emotions clearly have a role in these phenomena; what can biology tell us about them?

Emotions are associated with particular areas of the brain. For example, specific parts of the hypothalamus, a small organ near the middle of the brain, are directly involved in pleasure and aversion, while another organ of the brain, the amygdala, is heavily involved in the emotion of fear. But although the causes of emotions might be understood, the mechanism by which we become aware of them remains a total mystery. And the same goes for every other aspect of consciousness.

Some philosophers and neuroscientists cling to a dualistic interpretation of consciousness, claiming that it is separate from the physical world, and will therefore never be adequately explained by scientific theories.

fear factor
Hormones— chemical messengers coursing around our blood—are involved in our emotional state. Adrenaline can make us anxious, ready to fight or to run for our lives.

Some have attempted to locate consciousness in specific physical regions of the brain, including the reticular formation, which acts like a sorting office organizing sensory information from the whole body. Others consider consciousness to be an "emergent" property, not located in any particular brain region, but arising from the distributed activity of all the neurons. In particular, they point to the cortex—the incredibly complex, convoluted outer layer of the brain. This approach makes some sense, since only animals that have brains with a well-developed cortex seem to show signs of consciousness.

Our understanding of human intelligence has progressed enormously over the past hundred years, and the processing power of computers has increased at a phenomenal rate. However, despite all the investigations of philosophers, neuroscientists, and evolutionary biologists, the mystery of how we are able to feel aware of anything is a major hurdle for artificial intelligence researchers hoping to manufacture truly intelligent machines.

brainy beasts
Only animals with large brains, such as dolphins and chimpanzees, exhibit anything like human intelligence.

organ of awareness?
The reticular formation has been shown to route sensory information to the relevant part of the brain, and is most active when we are most aware. Could it be responsible for consciousness?

approaches to artificial intelligence

Despite our progress in understanding the human brain, the goal of building machines with minds of their own seems as distant as ever. There are two main approaches to building intelligent computers. The "top-down" approach attempts to model the workings of the human mind; top-down systems use symbolic representations of "concepts" and the relationships between them, and are successful in applications where logic and language are important. Bottom-up systems try to imitate the human brain, and are based on the way neurons work: networks of artificial, electronic neurons can learn tasks without being programmed to do so. Bottom-up researchers hope that their approach will ultimately result in machines that can think for themselves. Maybe the most elusive property of the brain—consciousness—will emerge from their systems.

mindless worker
Robots that carry out tasks on production lines are examples of top-down artificial intelligence. They follow inflexible programs and do not think for themselves.

top-down a.i.

top-down game
By analyzing the rules of tic-tac-toe, it is possible to identify strategies that always guarantee at least a draw. These strategies can be formalized into a computer program so that a computer can appear to play intelligently.

In January 1956, computer scientist Herbert Simon began one of his lectures with a remarkable statement: "Over Christmas, Allen Newell and I invented a thinking machine." Simon and Newell had written a computer program, which they called Logic Theorist, that could work out proofs of mathematical theorems by a process of logical deduction. Logic Theorist incorporated a set of rules and instructions—an algorithm—that used mathematical facts as a starting point and automatically deduced a number of fundamental mathematical proofs. Programmed deduction and decision-making make top-down systems very good at tasks that require logical reasoning, such as playing games, but not so good at tasks that involve flexible learning.

problem-solving

Within a year, Simon and Newell, together with another computer scientist, Cliff Shaw, had written a more general program—the General Problem Solver. Embedded within General Problem Solver's algorithms were "heuristics"—generalized approaches to problem solving—developed by analyzing human problem-solving techniques. So, for example, General Problem Solver could quickly solve logical games, such as tic-tac-toe and the Tower of Hanoi (below). In order to carry out these tasks, General Problem Solver used symbolic representations of facts, numbers, and even concepts.

In a computer, numbers, words, sounds, and images, are symbolized by "on" and "off" pulses—binary numbers. Computer programs contain instructions for manipulating these representations, which is why computers can process digital music and video and do arithmetic. Each program is written specifically for a particular task. An adding program contains different instructions from a program that sorts names into alphabetical order—just as the instructions you would give to someone to send a letter would be different from those telling someone how to travel to your home.

The US computer scientist **Allen Newell** (1927–1992) was working at the Rand Corporation when he met US economist **Herbert Simon** (1916–2001), who was lecturing in industrial administration at the Carnegie Institute of Technology. Together, they made important early advances in translating human problem-solving techniques into computer programs.

Computer programs make use of conditional statements, such as "IF ... THEN," which cause a computer to carry out different sections of a program according to the input they receive. This is why the top-down approach is suited to decision-making and deduction. With conditional statements, a program can respond to different kinds of input in an intelligent way. Part of a program might include an instruction to output "I am okay, thank you" if it receives an input of "How are you?" If you make the program incredibly complex, you can expect some very complex "behaviors" as output.

towering success
The top-down approach to artificial intelligence is successful at solving logical puzzles such as the Tower of Hanoi. The object of the puzzle is to move all the rings from one peg to another, one at a time, using the minimum number of moves, without ever placing a ring on top of one that is smaller than itself.

top-down artificial intelligence

Perhaps the best illustration of the top-down approach to artificial intelligence is a computer program that plays chess. When it is the computer's turn, it has to decide what move to make based on the positions of the chess pieces on the board. For each of its possible moves, there will be many moves that its opponent can make, and for each of these, there will be many more that the computer can choose from. So, the computer program "looks ahead" many moves, calculating which of its next moves is most likely to put it into a winning position. It does not care if it wins or loses: it merely follows a complicated algorithm. This built-in logic is the essence of the top-down approach. A chess program running on an ordinary desktop computer can analyze thousands of potential moves every second, and can beat most people easily. Very powerful computers can look much further ahead, and can beat even the world's best players.

decision tree

A chess computer uses its computational power to analyze all possible states of a chess board many moves ahead. In May 1997, a supercomputer named Deep Blue beat the world's best human chess player, Gary Kasparov. Deep Blue can analyze more than 200 million positions every second.

cognitive science

The human mind can be compared to a computer program—running on biological "wetware" rather than electronic hardware. Cognitive science is the study of mental processes—an investigation into how the mind enables us to think and communicate, and how it dictates human behavior. It can be thought of as a search for "computational" theories of the human mind. Many of the early pioneers of artificial intelligence were also cognitive scientists, and the development of cognitive science and computer science are intimately linked.

mind program
Early cognitive science was based on the idea that the brain is an extremely complicated computer. Nowadays, cognitive science is a wide discipline that is still very relevant to studies of artificial intelligence.

a grammatical understanding

One of the most influential cognitive scientists is US linguist and political activist Noam Chomsky. In 1957, Chomsky put forward a theory suggesting that that human brains are "preprogrammed" to understand grammar—the logical structure of a language. By building grammar into computer programs, top-down researchers hoped that computers would understand sentences, translate them from one language to any other, and even generate them. This area of top-down artificial intelligence research is called natural language processing.

US linguist and political activist **Noam Chomsky** (born 1928) is most famous for his radical theories proposing that language and several other facets of human intelligence are innate. He has often voiced radical views in politics, too—most notably during the 1960s, when he spoke out against US military involvement in Vietnam.

speaking the language
Real-time machine translation has found some uses: for example, the One-way Phrase Translation System, developed by the US defense research agency DARPA, can recognize simple sentences spoken by medical staff and play prerecorded translations of them in almost any language. It was designed to be used in battlefield and disaster situations.

During the 1950s, several attempts were made to enable machines to carry out translation. Many of the early attempts were crude, slow, and inaccurate. Even in 1966, automatic translation was making little headway: a report by the Automatic Language Processing Advisory Committee in the US condemned it as more expensive and not as effective as human translation. Today, computers are used in the translation of written documents, although human experts are still the first choice in most situations. And despite the fact that many computer software packages are available that can produce text from spoken words, human interpreters are almost always used to translate speech. In addition to translation programs, there are also several commercial products available today that can summarize long documents, by teasing out their main points. These are examples of top-down artificial intelligence.

and how does that make you feel?

conversational computers

Another facet of natural language processing is the attempt to create computer programs that can hold conversations with humans. The first of these programs was ELIZA, created in 1966 by Joseph Weizenbaum, a computer scientist working at the Massachusetts Institute of Technology. It was designed to mimic a "nondirective psychotherapist," responding to statements input by its "patients." ELIZA created a huge amount of public interest in artificial intelligence. Many people who interacted with the program formed emotional bonds with it, and a few even believed they were really relating to a human being. However, all ELIZA could do was to analyze the patients' statements and produce questions based directly on them.

natural language processing

All sentences can be broken down into separate phrases, and each phrase into smaller parts, and smaller again, until it is possible to identify the function of each word, and "make sense" of the sentence. Quite how the human brain carries out this process—called parsing—is not known, but it is possible to program a computer to do it. One popular application of natural language processing is the chatbot—a program that can converse with people.

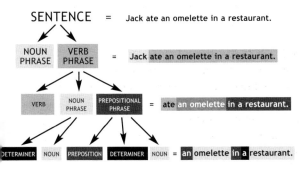

You can interact with chatbots on the internet, by typing in sentences and waiting for a reply. There are even some commercial websites that use chatbots to respond to users' questions about a company's products. As computers become more powerful, and as the internet

key points

• Most of the early successes in artificial intelligence were a result of the top-down approach

• The top-down approach is based on logical rules

• Cognitive science is an attempt to find computational theories of the human mind

parsing a sentence
A computer can be programmed to identify which words in a sentence have which functions, as shown here. Parsing a sentence is not the same as understanding it, however, and despite decades of research, programmed natural language processing is still a poor imitation of human language capabilities.

digital chat
In this transcription of a conversation with a chatbot, the sentences on the right are the computer's replies to an internet user's questions.

is used by more people, this kind of chatbot is likely to become more common. And as speech recognition and speech generation become more sophisticated, interacting with chatbots may become like human interaction.

a virtual world

Top-down computer programs that can effectively parse sentences and construct responses work best of all when the conversation is limited in its extent. The best example of this is SHRDLU, a program written in 1970 by computer science student Terry Winograd. It was written, as part of his PhD dissertation, as a way for humans to communicate in a natural manner with robots carrying out manual tasks. Users sat in front of a computer terminal that displayed a simulated "world" on the screen. The virtual world of SHRDLU consisted of a robot arm, and a variety of objects— simple geometric shapes—that could be moved on a flat surface. Because SHRDLU's virtual world was so limited, the responses it was able to give seemed extremely intelligent.

What is it like to be a chatbot?

A chatbot is a program like me, which you can chat with in plain English.

But do you like being a chatbot?

A chatbot is a program like me, which you can chat with in plain English.

Are you happy?

I have no idea.

What do you do when you are not online?

I love chatting on the internet!

What else do you do, besides chatting?

Chatting?

Anything else?

I give up.

Do you have any memories?

I'm sorry, I'm still being trained.

shrdlu

SHRDLU was an attempt to marry the reasoning powers of problem-solving programs with the natural language-processing capabilities of programs like ELIZA. To this end, the program consisted of: a parser, to deconstruct the sentences; a semantic component, designed to analyze the meaning of the deconstructed sentences; a "cognitive deduction unit" that could make logical inferences about the virtual world; and an English response generation component, to construct the responses. Here is part of the transcript of a demonstration of SHRDLU's capabilities.

Person: "Pick up a big red block."
Computer: "OK." (does it)
Person: "Grasp the pyramid."
Computer: "I don't understand which pyramid you mean."
Person: "Find a block which is taller than the one you are holding and put it into the box."
Computer: "By 'it', I assume you mean the block which is taller than the one I am holding."

intelligent responses
The SHRDLU program allowed operators to type in questions and receive correct answers. They could also type in instructions, and SHRDLU would update the virtual world.

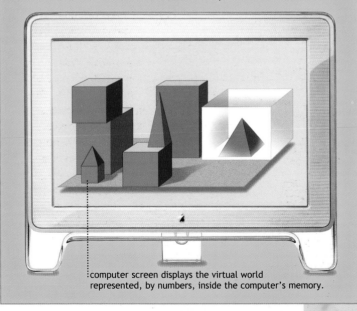

computer screen displays the virtual world represented, by numbers, inside the computer's memory.

expert systems

recap

The **top-down** approach is particularly good for solving logical puzzles and natural language processing, and forms the basis of expert systems.

intelligent help
Staff working on telephone support desks increasingly use expert systems to help them deal with callers' problems more efficiently.

Today, the deductive capacities of top-down artificial intelligence are most commonly applied in expert systems—computer programs that can analyze a stored database of information. The output can be information, advice, predictions, or risk assessment. One of the earliest and best-known expert systems was MYCIN, written by Edward Shortliffe in 1976 at Stanford University in California. MYCIN swiftly generated advice on how to treat bacterial infections of the blood, based on question-and-answer sessions with doctors about visible symptoms. Without MYCIN, most of these diagnoses would have required the culturing of bacteria from blood samples, a process that takes at least 48 hours.

Developments in the speed and storage capacity of computers have enabled designers of expert systems to increase computers' knowledge bases and increase the effectiveness of their programs. A project called CYC has a knowledge base of more than a million items, built up since 1984. There are many commercial applications of the CYC knowledge base: "Smart" interfaces between humans and computer databases; intelligent character simulation for games; improved translation; improved speech recognition. But part of the collection of "common sense" statements is also available to the public, via the internet. There are also projects that depend on the internet to increase the knowledge bases of expert systems. One is a program that can play a game of "20 Questions" in which a person thinks of an object that another person must guess by asking simple questions. The computer program adds to its knowledge base—according to whether it guesses right or wrong—with each game that it plays.

a logical step

Translating and summarizing documents, holding conversations, giving advice to doctors, and playing guessing games are all activities that seem to require genuine understanding. In top-down programs, they are accomplished by devising sequences of logical instructions. The world's foremost mathematicians have pondered the nature of logic since the middle of the 19th century. In an attempt to formalize all there is to know about logic—and about numbers, shapes, and space itself—they constructed complex mathematical proofs. In fact, it was a selection of these proofs that Herbert Simon and Alan Newell's Logic Theorist worked out automatically in 1956.

> **"You can only find truth with logic if you have already found truth without it."**
> G.K. Chesterton (1874–1936)

Had it been successful, the complete formalization of the world by mathematical proofs would have given weight to the idea that the brain can be simulated by a machine, since computers work according to the rules of logic. But the mathematicians' quest was dealt a serious blow in 1931 when a German mathematician named Kurt Gödel published a proof of his "Incompleteness

biological expert
Naturalists can use expert systems loaded into laptop computers to help them identify a species of plant or animal quickly and correctly in the field, as if they were carrying a human expert around with them.

Theorem," which proposed that the quest for a complete, all-encompassing theory was futile.

One of the consequences of Gödel's theory is that logical reasoning cannot be fully formalized; there always needs to be some "insight" in understanding any logical system, which must come from outside the system. In a comprehensive dictionary, for example, each word has a definition, but only in terms of other words in the dictionary. Users must apply their understanding of at least some of the words that are present in order to make any sense of those words that they do not. Closed systems, in which all the necessary insight is built-in, can be interpreted with the rules of logic. This explains why SHRDLU was able to communicate successfully and "understand" commands and questions. But "open systems"—real world situations—cannot be formalized, which explains why chatbots are not so successful in interacting with humans. There are many researchers who believe that the only way to construct a form of genuine intelligence is to allow a system to "experience" and learn the world directly for itself.

math is not enough
Try as they might, mathematicians cannot describe the world completely using mathematical logic alone. This poses a problem for logic-based attempts at artificial intelligence.

The work of Austrian-born US mathematician **Kurt Gödel** (1906–1978) on the "Incompleteness Theorem" is among the most important of the 20th century. In 1939, he fled Austria for the US, where he lived until his death. He starved himself to death after he became paranoid that he was being poisoned.

bottom-up approach

Somehow, the millions of signals from nerve endings in our sense organs form an internal representation of our immediate surroundings: our perception of the world. That internal representation relates to the firing of groups of neurons inside our brain. Learning, too, seems to arise from the formation of new connections between neurons—or the strengthening or weakening of existing ones. These connections are synapses, the tiny gaps between the axon of one neuron and the dendrite of another, across which the signals pass (see pp. 18–19). The bottom-up approach mimics these aspects of human intelligence, by creating networks of artificial neurons, each one behaving like a real, biological neuron.

copying nature

In 1943, neuroscientist Warren McCulloch and mathematician Walter Pitts were the first to design an electronic circuit that would behave in the same way as a biological neuron. The MP (McCulloch-Pitts) neuron had several inputs and one output. A real neuron has many "inputs," either signals received by its dendrites from other neurons, across synapse gaps, or else signals from the outside world in the case of sensory neurons. Some of the inputs encourage the neuron to fire, while others inhibit it. Each input to the MP neuron was binary—either on or off—but some inputs contributed negatively, others positively. If enough input was present in total, the neuron would produce an output, equivalent to a real neuron "firing." McCulloch and Pitts' artificial neurons were designed to carry out logical operations, as one way

organism

cell

molecules

atoms

built bottom-up
Bonds between atoms build molecules, and interactions between molecules make larger structures such as cells. Bottom-up programs also produce complex behaviors from simple instructions.

artificial neurons

The behavior of an individual neuron is fairly easy to reproduce in a computer, since a biological neuron can be considered as a two-state device—"on" when it is firing and "off" when it is not. A neuron "decides" to fire or not fire depending on the inputs it receives from other neurons.

This, too, can easily be simulated. The likelihood of a neuron firing is not related simply to the number of inputs: signals at some synapses act to increase this likelihood, while at others they reduce it. Even this can easily be replicated in a computer or electronic circuit to form "artificial neurons."

artificial neuron

INPUT 1 x WEIGHT 1

INPUT 2 x WEIGHT 2

INPUT 3 x WEIGHT 3

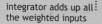

OUTPUT

integrator adds up all the weighted inputs

threshold unit produces output if the sum of the weighted inputs is greater than a certain value

biological neuron

axon carries "output" pulses

cell body

output (firing or not firing)

dendrites pick up signals from the axons of other neurons

mimicking nature
Just like biological neurons, artificial neurons have a single output and many inputs. The input weight of an artificial neuron corresponds to the efficiency of the synapse

to construct a digital computer. When it was shown that the same logical operations could be carried out by simpler circuits, research into artificial neurons was neglected.

new perceptions

In the late 1950s, Canadian psychologist Donald Hebb explained how neurons are involved in learning. He realized that learning takes place when the synaptic connections changed—as a result of repeated sensory input. So, for example, we learn to recognize a particular person's face by seeing it often, a process that results in changes in the synaptic connections between neurons in our brains. A US computer scientist, Frank Rosenblatt, adopted Hebb's ideas, and designed artificial neurons that could be made to learn. Rosenblatt's invention—the Perceptron—was modelled on human vision. The Perceptron was connected to the output of a light-sensitive cell, and the signals from the photocell were fed to a small, interconnected network of artificial neurons. After being exposed to many examples of letters of the alphabet, it could learn to distinguish them. The Perceptron was carefully trained to learn new patterns, by being shown several examples of an image. The output of the network was its interpretation of each particular image. Any errors in its interpretation were fed back into the network: the weights of the inputs to the neurons gradually, but automatically, changed according to the degree of error.

During the 1960s, the Perceptron enjoyed a good deal of attention, and large grants were given for research into its potential applications. Several military applications were proposed—could a Perceptron be trained to recognize enemy

digital image
Computers work only with numbers. Images and sounds must therefore first be represented as numbers, or "digitized," before they can be processed by neural networks. A digital camera does just that.

A =A
A =A
ℰ =A
A =A
A =A
A =A
A =A
A =A

any which way
A suitably trained Perceptron can identify which letter of the alphabet is being input as a digitized image—even if the image differs from those used in its training.

artificial neural networks

In the brain, it is the interconnection of many neurons that makes us able to perceive, learn, and remember. These aspects of intelligence can be created by combining many artificial neurons together—either as an extremely complicated computer program or by physically connecting electronic circuits that behave like neurons—to form artificial neural networks.

These neural networks are usually arranged in successive layers, with the output of each layer forming the input of the next. Just like networks of biological neurons, artificial ones have inputs from the outside world—each artificial neuron in the input layer has only one input, just as each

number stream
When the image falls on the array, a stream of numbers is produced. These numbers represent the brightness of the image at each element.

bright eyes
The numeric information about the brightness of the image at each point on the array can then be sent to a computer.

lightly focused
A lens collects light from a scene and focuses it onto an array of light-sensitive elements called a CCD (charge-coupled device).

sensory neuron in our body has
one input. The inputs to an
artificial neural network that learns
to recognize images are the values
representing the brightness of
each pixel of an image; the inputs
to a network that finds patterns
in large databases of information
are the individual pieces of
information, expressed
as numbers.

image or input
*Inside a computer, the numbers making
up an image can be used to reconstruct it
on-screen—or they can be sent to the input
layer of an artificial neural network, which
could learn to recognize an individual face.*

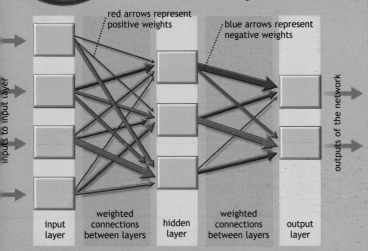

red arrows represent
positive weights

blue arrows represent
negative weights

inputs to input layer

outputs of the network

| input layer | weighted connections between layers | hidden layer | weighted connections between layers | output layer |

a simple artificial neural network
*Interconnected artificial neurons are arranged in layers. Neurons in each layer except
the input layer have weights associated with them. The inputs to each neuron are
multiplied by the weight, and the weights change as the network recognizes patterns.*

vehicles or spot patterns in satellite images, for example? However, in 1969, the bottom-up approach to artificial intelligence suffered a major setback, with the publication of a scientific paper about the theory behind Perceptrons. The paper showed that there are some simple logical operations that cannot be carried out using a Perceptron, but which can be carried out using a conventional

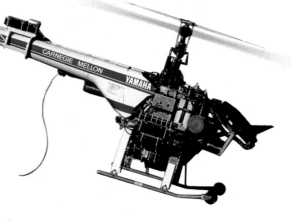

finding its way
Back propagation neural networks have been successfully employed in a wide range of applications, including self-driving cars and self-flying helicopters.

computer. This news had the effect of halting nearly all work on artificial neural networks for about 10 years.

One response was to make the Perceptron more complicated, by incorporating a second, followed by a third "layer" of artificial neurons. Each "hidden" layer of neurons is fed with the output of the neurons in the previous layer. On its own, this solves nothing: these multi-layered Perceptrons suffered from the same limitations that their predecessor had. However, feeding the errors back through each layer of the artificial network—rather than just to the input neurons— extended the range of computational tasks that it was able to carry out successfully. This "back propagation" has the effect of improving the learning capability of a network, and the input weights of the neurons in the intermediate layers form a kind of short-term memory.

learning backward

Back propagation networks are commonplace in many applications of artificial intelligence. They find a use mainly in applications where pattern recognition is required. For example, some astronomers use these networks to help them classify distant galaxies; researchers have used them in face- and vehicle-recognition systems; financial traders use neural networks that can spot trends in stock market trading, to predict where to invest their money. Neural networks can be made to control self-driving cars, by monitoring the edges of their lanes and recognizing potential hazards.

Back propagation neural networks can be simulated on an ordinary computer. This is convenient: you can buy programs that run on a personal computer, rather than having electronic circuits custom-made. However, neural networks are also manufactured as stand-alone components—where a custom-built network-on-a-chip is cheaper. In stand-alone networks, the individual neurons work at the same time, in parallel, rather than one after the other. This parallel, or "distributed," processing more closely mimics the brain, which does not seem to have a central processor that carry out calculations one at a time.

who's that?
Facial recognition systems use neural networks processing digitized images. Typically, the network analyzes small sections of the images, searching for parts that look like elements of faces.

stand-alone network
The 3-Dimensional Artificial Neural Network (3DANN) is capable of recognizing objects in real time and in highly cluttered settings. It was developed as an artificial vision system for spacecraft and robots.

3DANN has layers of artificial neurons connected in parallel.

learning algorithms

Back propagation neural networks really can learn, but they bring us no closer to building an electronic brain. The fact that they can be simulated on an ordinary computer leaves them open to the same criticisms and subject to the same limitations as the top-down approach—in other words, they will lack insight, because they are formalized systems. Also, real neurons do not use back propagation. In the brain, signals only flow from the axon of one neuron to the dendrites of another, not in the reverse direction. Another fact that sets back propagation neural networks apart from biological ones is the supervision needed during learning. You have to train a network carefully and very specifically to make it learn what you want it to learn. This is not quite natural development, or genuine intelligence. Finally, the artificial neurons of which these networks are composed are oversimplified: a neuron in your brain is not really a two-state device like the neurons in artificial networks. A real neuron is "off" when it is firing infrequently, and "on" when it is firing rapidly, but it can be "partly" on, too—there is a range of firing rates.

These problems have been addressed in many ways. For example, new network "architectures" are employed in the hope that artificial neurons can learn more effectively. Another solution is to employ "fuzzy logic." Unlike the simple binary (two-state) mathematics used

do-it-yourself
Computer programs that allow computer users to design their own artificial neural networks, by defining the weighted connections between artificial neurons, are readily available. The result, however, is an algorithm: no different in principle from the top-down approach.

inside most computers, fuzzy logic works with a range of values. In effect, this allows artificial neurons to be partly on, as well as "on" or "off." Fuzzy logic has been used where "common sense judgment" is needed. Fuzzy controllers in electronic washing machines, for example, can be used to assess the state of the laundry to reduce the amount of water and electricity used in each wash.

Another "humanizing" approach to artificial neural networks is to reduce the amount of training required. Unsupervised learning by networks—self-organization— can find relationships within huge amounts of information which people would never be able to identify. One type of self-organizing network is the Bayesian network. Based on a theory by 18th-century British mathematician Thomas Bayes, Bayesian neural networks infer patterns even in situations where large portions of information are missing. Bayesian networks make educated guesses and common-sense predictions about a body of information, given previously known relationships about the kind of information present.

fuzzy thinking
Electronic circuits called fuzzy controllers in modern washing machines can decide how much water and heat to use according to various measurements.

living algorithms

A neural network can be represented as an algorithm—a sequence of instructions, which can be coded as a computer program. When the program and the neural network are run side by side with the same inputs, both systems will produce the same output. This is how neural networks are simulated on an ordinary computer.

incoming!
A Bayesian network tested by the US Navy can calculate the best defense against an incoming missile in just a few seconds.

A network that learns unsupervised defines its own best algorithm— the most successful solution. This is what happens in evolution, and some artificial neural networks are designed to mimic evolution. Two key features of biological evolution are the genome and natural selection. The genome is an organism's entire sequence of deoxyribonucleic acid (DNA), and there is a copy of it inside most cells of an organism. The DNA carries instructions on how to build that organism. All organisms in the same species have very similar genomes, but each organism's genome is slightly different. Sexual reproduction results in a mixing of DNA from two individual organisms of the same species, to create a new organism with another unique genome. Mutation—the introduction of small random changes in the DNA—can introduce new features into the genome, and therefore into the species. Some mutations work better than others, which leads to the best adapted, or "fittest," organisms surviving. This natural selection drives the development of the species.

Copying how natural selection works—but using algorithms instead of genomes—allows artificial neural networks genuinely to evolve. Researchers begin by devising algorithms that describe several different networks for a single problem.

darwin's finches
British naturalist Charles Darwin formulated his theory of evolution after an expedition to the Galapagos. His theory accounted for variations in species of Galapagos finches.

learning to walk
Unprogrammed robots controlled by neural networks can quickly learn to walk effectively using genetic algorithms.

ROBUG
University of Portsmouth

artificial life

Genetic algorithms are the point at which artificial intelligence research meets another, related, area of science: artificial life. The main aim of artificial life research is to investigate human-made systems that have some of the essential properties of life. The focus of artificial life research lies in computer science, where many of the features of living systems can be simulated mathematically.

By programming a computer with simple rules, applied to many separate "cells," complex environments can "evolve" quite naturally, and interactions between the individual cells can take on lifelike qualities.

One of the best-known examples of artificial life is called the Game of Life. Invented by US mathematician, John Conway, the program involves a simple grid of cells that can "live" or "die" as the program runs according to a straightforward set of simple rules. Each application of the rules to all of the cells is called a generation. The appearance of the cells in the grid changes in surprising and unpredictable ways over successive generations.

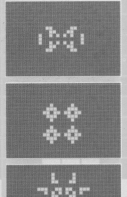

evolving forms
When the Game of Life is played out as a computer program, many interesting multicellular shapes emerge rapidly from the simple rules.

By "cross-breeding" the most successful algorithms, and introducing "mutations," artificial evolution takes place. Natural selection over many generations develops the ideal network algorithm for a particular goal. That goal could be to control a robot to walk efficiently with six legs, or for a robot with wings to hover. These real-world goals lead to an impressive array of lifelike behaviors that mirror the solutions that biological evolution has created. So, a six-legged robot quickly learns to walk in the same way as a real insect. The field of robotics is where "genetic algorithms" are applied most fruitfully.

emergent behaviors

Among the most exciting experiments in artificial intelligence are those in which human- or animal-like behaviors arise—or "emerge"—in robots or computer systems without being specifically programmed to do so. Particularly exciting are experiments involving teams of robots, each with limited computing power and simple behavioral goals, such as following each other and avoiding obstacles. Each robot is equipped with sensors, such as simple devices that detect the positions of nearby objects or pick up signals from the other robots. Behaviors that have been observed to emerge in such experiments include flocking, predator-prey behaviors, and self-imposed leadership. Each robot acts autonomously but the team works together in the same way as a colony of insects or a flock of birds. There are several potential advantages of such systems: they are adaptable to changing conditions, and can withstand the loss of one or more of the team.

meeting of simple minds
In some experiments of this type, the robots exchange flashes of light between each other, a simple form of communication. This is not programme into the robots, but emerges as a best way to achieve the collective goal.

random beginning
In a typical experiment into emergent, collective behavior, a number of small robots are placed randomly and given a simple goal—in this case, to move in a particular direction together.

a sonar system is used to avoid, or follow, solid objects

robot has limited computing power

team members
Robots taking part in experiments into emergent behavior are typically small, mobile, and with a set of simple sensors. In the future, fleets of self-driving vehicles might be used in real-world applications of the phenomenon, such as planetary exploration or battle maneuvers.

follow the leader
Often, a leader emerges, and the other robots follow it. Again, this is emergent behavior, not programmed into the system.

getting there
As the experiment proceeds, the robots achieve their goal of moving together in a particular direction— despite the fact that each one is working autonomously and has limited computing power.

intelligent robots

roam alone
Planetary rovers require a high degree of autonomy, as radio signals can take several seconds or even minutes to pass between Earth and the planet they are exploring.

The term "robot" originates in a play written in 1919, by Czech author Karel Capek, called "Rossum's Universal Robots." The word "robot" comes from the Czech word *robota*, used to describe work that is boring and repetitive.

Robots are machines controlled by computers, and their most common use in today's world is as a replacement for human factory workers, carrying out complex tasks that require great accuracy but are routine. Manufacturing robots therefore require only very limited intelligence.

One of the best-known humanoid robots—**Cog** ("born" 1994)—has been designed to interact in human ways, in an attempt to develop humanlike elements of intelligence. It has two video cameras for eyes, with which it follows moving objects. It stands on a pedestal, has two arms (the second added in 1997), and its brain is in a cupboard in the room next to where it stands.

In certain applications, however, such as planetary exploration, robots need to respond to changing circumstances autonomously—without direct human control. The bottom-up approach to artificial intelligence is well suited to creating such autonomous, semi-intelligent robots, typically controlled by onboard artificial neural networks that can recognize objects or sounds, adapt to their surroundings, and learn from experience. Perhaps the most

probe tests atmospheric gases

solar panel generates electricity

a computer partly controls rover

impressive of today's robots are humanoid robots, so called because they resemble human form. They have "eyes," "ears," and "bodies," and can learn a range of complex behaviors without being shown. The most famous of these is Cog, one of the many robots developing their own behaviors at the Massachusetts Institute of Technology. Cog has learned to move its eyes swiftly to shift the focus of its attention, as occurs in human vision. Cog has also taught itself to play the drums, to imitate the head-nodding of a person standing in front of it, and a range of other behaviors. The Kawato Dynamic Brain Project, in Japan, has yielded very similar results.

human form
The H5, built at Tokyo University, is one of several ambitious humanoid robot projects. The robot has learned walking, juggling, and many other human abilities.

a desirable future?

Attempts to create artificial intelligence have already made impressive progress. Expert systems and neural networks are becoming more widespread and more powerful, for example. But most of the commonplace applications of artificial intelligence are simply a particular approach to writing computer programs, and the dream of conversing naturally with a computer, as we do with each other, still seems distant. Can top-down systems become so sophisticated that they can exhibit real intelligence? Or is the bottom-up approach the only way forward—allowing intelligence to develop naturally as it seems to do in human beings? What might the fruits of artificial intelligence be, and do we really want to build machines that can think for themselves?

the future of artificial intelligence?

The idea that inanimate objects can be given life and human powers of intelligence or consciousness dates from long before the age of electronic computers—in myths, legend, and fiction. Perhaps genuine artificial intelligence is just a modern version of a recurring myth. Maybe it only occurs in science fiction. The philosophical arguments surrounding Alan Turing's Imitation Game, John Searle's Chinese Room Argument and Gödel's Theory of Incompleteness are deep and, for many people, unresolved. Will human intelligence be simulated, replicated, or even surpassed? As computer technology continues to grow in speed and sophistication, what will the future hold for artificial intelligence? Might future generations converse with computers as today we can only do with each other? Will computers one day be artists, politicians, and teachers? Or will they always be mindless electronic servants?

future fantasy
Artificial intelligence features heavily in science fiction films, such as Artificial Intelligence: AI, *directed by Stephen Spielberg, in which a robot boy longs for real love from his adopted human mother.*

intelligent agents

One setting in which artificial intelligence will feature increasingly is the internet. Many of the features that make the world wide web, the growing collection of interlinked "pages" of information available to internet users, useful are also time-consuming and laborious. Tasks such as searching for a particular item of information, or keeping abreast of developments in a specific field of interest, can be automated. Computer programs called intelligent agents can trawl the world wide web, searching out content that is relevant to a user's needs; some can learn a particular internet user's interests, and download relevant content.

mental legwork
Many of the tasks involved in using digital networks such as the internet are time-consuming and require only limited intelligence. Intelligent software agents are therefore being employed to take over these tasks, leaving only the decisions that require real intelligence and insight to human computer users.

The most useful intelligent agents behave intuitively. In the business sector, intelligent agents are already used to carry out financial transactions with very little human input—finding the best deal and communicating with other intelligent agents to work out the conditions of sale. Others automatically "read" online newspapers and journals, finding information relevant to a particular company and paraphrasing it. In the future, intelligent agents may offer people a more human interaction with their computers, by recognizing and generating speech.

electronic brains

The bottom-up approach to artificial intelligence has been successful in producing computer systems and robots that exhibit learning and pattern recognition. However, by simulating the way in which our minds process information, the top-down approach is successful in tasks that require natural language processing or deductive reasoning. Some researchers believe that, given time, even these abilities will also emerge in bottom-up systems—just as learning and pattern recognition have done.

Inside a computer in Tel Aviv, Israel, is an evolving computer program called Hal. Modeled on a human infant, Hal is learning English, slowly but surely. After two years, the program was able to "communicate" in sentences reminiscent of those produced by a toddler: three- or four-word phrases that are generated by association with previous experience. Child development experts reading transcripts of Hal's conversations considered the conversations to be typical of a human infant. Hal's creators believe that language is a learned ability, and employ a system of rewards and punishments to help train their electronic infant. They hope that by learning language from the bottom up, they can instill within the system a real understanding, missing from traditional natural language processing, which imposes understanding from the top down (see p. 31).

" We should not pretend to understand the world only by the intellect. The judgment of the intellect is only part of the truth. "
Carl Jung (1875–1961)

electronic infant
The approach taken by the creators of Hal is modeled on the constructivist approach to education, which argues that learning takes place inside the learner, rather than being imposed from the teacher.

bigger and better

Another approach to genuine intelligence is to build faster, bigger computers with more artificial neurons interacting with each other. In 1993, a team of researchers

Australian-British computer scientist **Hugo de Garis** (born 1947) has a long and impressive pedigree in artificial intelligence. Many experts in his field consider him a maverick with unrealistic visions, but de Garis is a fervent believer in the bottom-up approach to artificial intelligence. He predicts that thinking machines will become a reality in the foreseeable future.

based in Japan set out to design an electronic brain with millions of artificial neurons. The Cellular Automata Machine (CAM) was to be used to control a robot kitten. In 1997, the project team contracted a company in the US to construct the CAM Brain Machine (CBM). Completed in 2000, it consisted of more than 74 million artificial neurons, and its processing power was equivalent to 10,000 desktop personal computers. The individual neurons in the CBM were built on electronic circuits called field programmable gate arrays (FPGAs). These components can be constantly and quickly "reprogrammed"—the equivalent of altering the connections between neurons in the brain. The neurons of the CBM could be updated 300 times every second by genetic algorithms and could, it was hoped, evolve lifelike behaviors.

The CBM has never been connected to a robot kitten, as originally planned, and, despite the resources of its huge computational powers, it has not yielded anything remotely resembling human intelligence, although it has been used in general purpose supercomputing, in business and scientific research. Many artificial intelligence researchers nevertheless believe that the CBM is too simplistic in its design. It is better, they say, to design bottom-up systems that address the structure of the human brain more closely.

robot kitten
A virtual kitten was to be the interface between humans and CBM, as it was hoped that the machine would have the same intelligence as a cat. But in 2002, this project was scrapped.

closer to reality

There are many bottom-up projects that focus on specific features of the human brain to bring us closer to creating genuine intelligence. One aspect of biological neurons that is not generally included in the design of artificial neural networks is the action of neurotransmitters—molecules that carry nerve signals across the synaptic gap between neurons. Neurotransmitters are released at the ends of axons whenever a neuron fires. When they reach the other side of the synapse, they are absorbed by the receiving neuron. According to conventional neuroscience, neurotransmitters are all relatively large molecules, which can only travel short distances. However, during the 1990s, neuroscientists discovered that some neurons emit neurotransmitters whose molecules are smaller, and which can therefore "diffuse," over a relatively large area, affecting hundreds of other neurons.

gas injection
By simulating the subtle features of the human brain, such as the diffusion of the neurotransmitter nitric oxide, researchers are able to increase the efficiency of their artificial neural networks.

virtual gas

Nitric oxide has been shown to modulate the firings of neurons that lie within its range of diffusion; the amount of modulation depends on the concentration of the diffusing nitric oxide. Phil Husbands of the University of Sussex, England, simulated diffusable, modulatory neurotransmitters in his experiments with neural networks. He used genetic algorithms to develop a network which, when connected to a video camera, could recognize a triangle. The effect of nitric oxide was simulated on a mathematical model, and this "virtual gas" diffused through the network. Using the virtual gas, the network evolved about 10 times faster than a network using genetic algorithms, and needed fewer artificial neurons to complete the same task.

the subtle brain

Despite the exciting achievements of bottom-up artificial intelligence researchers, most would agree that their artificial neural networks are simplistic, that there is much more to human intelligence than electrical pulses and changing synaptic connections. However, our understanding of the brain is increasing every year. It was only in the 1990s that neuroscientists began to realize the role of nitric oxide is to act as a neurotransmitter. Unlike other neurotransmitters, it can affect all the neurons within its diffusion range, not just those to which it connects. There are without doubt many other phenomena of which we are currently unaware, but which contribute to our intelligence. Until we understand all the brain's subtle but important features, we will probably remain a long way from creating true intelligence in a machine.

network of neurons in the brain

nitric oxide diffuses, affecting synapses within this area

nitric oxide is released at a synapse

spreading influence
Nitric oxide diffuses in all directions within the brain, affecting many neurons.

meeting in the middle

key points

• Intelligent agents will become more common
• Building more powerful computers does not create artificial brains with true intelligence
• Humanlike intelligence can only be created with an understanding of the human brain

The brain is not simply a collection of self-learning networks of neurons—it does have structure and order. For example, the brain has an amazing ability to focus its attention, finding order in the signals it receives from the outside world. Different types of neurons, in a variety of arrangements, seem to give us this ability. One theory suggests that information passes from parts of the brain dealing with "higher" processing into areas of the brain that deal with perception, forcing those areas to focus their attention on a particular aspect of the world. This higher information is fed through groups of neurons called pointer neurons which, when included in artificial neural networks, focus attention. They make networks able to recognize certain patterns when other patterns are more prominent.

Name: Joe Bloggs
Sex: Male
D.O.B: 12/18/68
Height: 198cm
Weight: 55kg
Job: Designer

neuromorphic engineering

Nearly all attempts at creating artificial intelligence are carried out with digital computers—machines that use binary (two-state) arithmetic to carry out logical operations (see p. 27). Digital electronics is the key to all-purpose computing, but it may not be the key to artificial intelligence—there is a promising alternative. The opposite of digital electronics is analog electronics, in which inputs and outputs are determined by the amount of electric current flowing, not by abstract representations of numbers as in digital

trained eyes
A neural network that includes pointer neurons could perhaps be able to distinguish the face of a missing person in a busy street scene.

electronics. In the emerging discipline of neuromorphic engineering, researchers create networks of analog electronic components that are modeled closely on structures in animals' brains. In this system, it is the actual measurements of electric current flowing that is important. Neuromorphic circuitry is simpler, faster, and uses less power than conventional digital circuits.

electronic senses

Most neuromorphic experiments have concentrated on building electronic senses. The best example is the retina—the light-sensitive "screen" inside the eye. Working with electronics guru Carver Mead, neuro-morphic engineer Misha Mahowald designed a "silicon retina" composed of groups of artificial, analog neurons that carried out the same filtering processes on images as neurons do in the human retina. Another neuromorphic approach to vision was based on equipping robots with the neuromorphic version of a housefly's complex visual system. The circuitry uses less power than earlier robot visual systems.

Neuromorphic engineering will have a huge impact on artificial intelligence research. As the circuitry imitates biological systems, neuro-morphic components may be implanted into the body, as replacements for faulty ears, eyes, or noses. Artificial sense organs will interact with biological neurons, sending messages to the brain. For this to happen, electronic circuits must be able to communicate effectively with biological neurons.

future vision
In the science fiction series "Star Trek: The Next Generation," Geordi LaForge was born blind and he wears a visor that gives him better-than-human vision. This may become possible in the future, as scientists develop neuromorphic devices that can communicate with biological neurons.

making the connection

Several experiments have already been carried out involving communication between electronic circuits and biological neurons. John Chapin and his colleagues at the MCP Hahnemann Medical College, in Philadelphia, implanted an array of electrodes into a specific part of a rat's brain. The team had trained the rat to press a lever when it wanted food, and they recorded the outputs of the neurons every time the rat pressed the lever. When the electrodes picked up the signals, a device delivered the food without the rat having to press the lever. The rat learned that all it had to do was "think" about pressing the lever to receive the food.

This ability to detect intention in the neurons of a brain holds potential for "intelligent" prostheses—artificial arms or legs, directed by thought power. An experiment at Emory University, Atlanta, involved people paralyzed from the neck down by spinal injury. The patients' own neurons grew into a small glass case containing a set of electrodes implanted in the brain. Weeks later, the patients' brains had learned to control a pointer on a monitor by sending signals, generated in the brain, through the electrodes.

mind control
Communication between biological and artificial neurons has already been established. Humans, as well as rats, may soon be able to control electronic devices by the power of thought alone.

nerve power
The neuronal transistor, developed by German biophysicist Peter Fromherz, allows two-way communication between an electronic circuit and a microscopic nerve cell, taken from a leech, bonded to the silicon.

hybrid neural networks

space pioneers
Robots require no life-support systems, enabling them to survive long journeys through space, requiring only electric power. Endowed with enough intelligence, robots may form a future fleet of deep space astronauts. If humanoid robots make it into space, they will not need to wear spacesuits.

In a related, and equally thought-provoking experiment, biological neurons have been made to attach to, and communicate with, miniature integrated circuits. The union of living neurons with silicon—the material from which integrated circuits are made—was problematic. One reason for this is the fact that the neurons grow and move around on a silicon surface, making the interface between the two materials difficult to maintain. Several different solutions were developed, including building tiny "picket fences" that hold the neurons in place. The development of hybrid neural networks, involving electronic circuits and real neurons, may eventually lead to implants that can be bonded directly into our nervous systems, perhaps to extend the capabilities of our brains.

the future of robotics

Future developments in artificial intelligence will certainly lead to robots that are more intelligent, more adaptable and more commonplace. Most of the exciting developments in robotics are likely to be in military applications and in space exploration. Already, semi-autonomous robots have explored part of the surface of Mars. With genuine intelligence, robots would be able to fly the entire mission, collect samples, and carry out experiments. Intelligent robots would also be heavily involved in constructing bases on the Moon or Mars— they would be able to survive the harsh conditions,

and could build shelters for human habitation. The military applications of robots are already wide ranging. Many guided missiles now make use of artificial intelligence to ensure that they hit their targets. Self-flying fighter planes and small flying surveillance robots are also under development. In the far future, the military applications of intelligent robots could be as dreadful as they are impressive.

electronic servants

Since the 1950s, it has been the dream of many people to have robots in their home, carrying out tasks such as household cleaning, and preparing and serving food. In the next few years, that dream may in part be realized. Several automated home-helpers are becoming available, including an "intelligent" vacuum cleaner that can avoid obstacles, and others will appear in the next few years. These robots will have very limited intelligence, but in the distant future, perhaps every home will have its own thinking machine. Robots could be personal assistants or even companions. To be true companions, robots would have to have minds of their own—but to create them requires a solution to the greatest mystery of neuroscience: consciousness.

home help
If artificial intelligence researchers can create genuine thinking machines, you may one day find an electronic servant bringing you the morning paper.

truly conscious?

silicon dreams
People imagine computers with minds of their own, but is it possible for computers to have hopes and dreams of their own?

Artificial neural networks have been observed doing something like dreaming—when their inputs are disconnected, the internal state of these networks generates its own patterns of activity. But the creation of consciousness remains a distant prospect for artificial intelligence research. Some researchers hope that it will emerge in their bottom-up artificial neural networks, just as some other behaviors have. But consciousness seems to be much more than just a behavior, and a study of consciousness still lies mainly within the realms of philosophy: no one knows what it is or how it works.

One recent theory, proposed by American medical researcher Stuart Hameroff and British mathematician Roger Penrose, suggests that tiny structural units within neurons, called microtubules, are heavily involved. Microtubules are long, incredibly slender cylindrical molecules that lie deep inside most of the body's cells—including neurons. It has often been suggested that quantum theory may be involved in consciousness in some way. According to Hameroff and Penrose's theory, information is passed between the tiny molecules that form the microtubules in the brain's neurons. The rules of quantum theory are beyond the scope of this book, but if Hameroff and Penrose's theory is correct,

"Computing is not about computers any more. It's about living.

Nicholas Negroponte

then it is almost certainly impossible to produce consciousness in any conventional computer. However, computer scientists are already examining the possibility of using quantum theory as the basis of information processing in future computers. Quantum computers would be far more powerful than conventional computers—might they also be conscious?

spiral structure of individual microtubule

microtubules

axon

synapse

dendrite

cell body

inner world
Microtubules are incredibly tiny molecular structures deep in human cells. Might they be the key to our ability to experience the world around us?

It seems reasonable to assume that when our understanding of intelligence is advanced enough, our efforts to mimic newly discovered features of the brain will take us closer to creating machines with true electronic brains.

friend or foe?
The pessimistic among us may be happier if machines with minds and free will never become a reality.

What happens if we create machines with an intelligence equivalent to our own? What if they become hungry for power and frustrated with the frailties of their human creators? Might they pose a threat to human life? If they ever exist—intelligent machines with independent thoughts, memories, desires, and understanding—would they be considered living beings? Would they have rights? Would it be murder if you turned one of them off? If developments in artificial intelligence continue, then these questions could soon become very relevant.

glossary

algorithm
A clearly defined, step-by-step procedure for solving a problem. For every mathematical problem that is solvable, there is an algorithm, and any algorithm can be written as a computer program.

analog
The opposite of "digital." Most computers use digital electronics, working with binary numbers that have "discrete" values—just as you can only stand at certain heights on stairs. Analog electronic circuits work with "continuous" values—just as you can stand at any height on a continuous slope.

artificial life
A branch of science that attempts to reproduce certain essential features of natural, living systems in nonliving ones, particularly computers. Artificial life and artificial intelligence are closely related. *See also* genetic algorithm.

artificial neural network
A collection of interconnected artificial neurons, which can be used to solve mathematical problems. Artificial neural networks can be manufactured as stand-alone electronic components, but can also be simulated as an algorithm on a computer.

artificial neuron
An electronic circuit that is designed to behave in a similar way to a real, biological neuron. An artificial neuron can also be simulated in a computer, as an algorithm.

autonomous
Describing something under its own control. One of the major goals of artificial intelligence research is to design autonomous robots.

axon
The branch of a biological neuron that carries the neuron's "output." A neuron has only one axon.

back propagation
A technique used in most artificial neural networks, in which the weights of the artificial neurons are changed by an amount that depends upon the degree of error in the output of the network. These errors are fed back through every layer of the network, allowing networks to learn to recognize new patterns very effectively.

bayesian network
A type of artificial neural network that learns to recognize patterns in large amounts of information.

Bayesian networks are used in a wide range of applications, and are among the most useful and most powerful artificial neural networks. *See also* self-organization.

binary
The number system that has only two digits (0 and 1), used as the basis of digital electronics at the heart of most computers.

bottom-up
The approach to artificial intelligence that attempts to mimic the architecture of the brain, using artificial neural networks. *See also* emergence.

chatbot
A computer program that can hold a conversation—typically via the Internet—with a computer user. It is able to analyze sentences and construct replies. Some chatbots learn from their conversations, making them appear convincingly human. *See also* turing test.

cognitive science
The branch of science concerned with developing theories about human perception, thinking, and learning.

CPU
Acronym for central processing unit; the part of a computer that carries out

the instructions that form computer programs. Inside a CPU, instructions and information are represented as binary numbers.

dendrite
The part of a biological neuron that receives stimulation from other neurons. Most neurons have hundreds of dendrites. *See also* synapse; axon.

dualism
The idea that the mind is separate from the physical world, and cannot be explained by scientific theories. Most neuroscientists believe that dualism is false.

emergence
The appearance of a behavior or some other feature of a complex system, typically a system made up of many semi-independent parts interacting together. Many neuroscientists believe that consciousness is an emergent property of the neurons that make up the human brain, and bottom-up artificial intelligence researchers hope that they may one day reproduce it in a computer.

expert system
A computer program that gives nonexperts in a particular field access to expert knowledge. Typically, an expert system has an organized collection of facts and concepts—called a knowledge base—and some way of interacting with and

updating it—the inference engine. Expert systems may be used to help people learn a new skill, to advise people, or to assist them in some other way.

fuzzy logic
A branch of mathematics and computer science in which "part truths" may be used. In most computing applications, a particular statement is either true or false according to the rules of conventional logic.

genetic algorithm
An approach to solving a mathematical problem that mimics natural evolution, used with artificial neural networks. To find the best algorithm for a particular problem, many possible algorithms are tried, and the best ones selected and changed slightly, then tried again. The process is repeated over many "generations."

heuristics
Generalized approaches to solving problems—"rules of thumb"—that can be used to design computer programs.

hidden layer
Any layer of artificial neurons that lies between the input layer and output layer of an artificial neural network.

intelligent agent
A semi-intelligent computer program that can carry out laborious or repetitive tasks, saving time for computer users. Intelligent agents are

typically found on the internet, where they carry out automated searches of the world wide web.

intentional state
Any "felt" internal state of mind that carries meaning or intention, including beliefs, desires, fears, and pain. Intention is a major hurdle in creating truly intelligent computer systems.

logic gate
An electronic circuit found in a computer that carries out simple logical operations using binary mathematics. Many interconnected logic gates are found inside the CPU (central processing unit) of a digital computer.

microtubules
Tiny protein structures found within all cells of living things. Some neuroscientists believe that microtubules are important in creating consciousness.

natural language processing
The process by which computers can analyze sentences and produce grammatically correct responses. *See also* parsing.

neural network
See artificial neural network

neuromorphic engineering
The branch of artificial intelligence which uses analog electronics instead of digital electronics, mimicking the behavior

of elements of the human brain more directly than other approaches.

neuron

The basic unit of the nervous system of an animal. The operation of a neuron is well understood, and most neuroscientists believe that the collective operation of millions of neurons in the human brain gives rise to learning, perception, and even consciousness.

neurotransmitter

Any of several chemicals produced by neurons in the brain, which carry nerve impulses across the gap between two neurons.

parallel processing

The approach to computing that uses many processing units (*see* CPU) acting at the same time, each one tackling part of a problem.

parsing

Breaking down a sentence into its component phrases, in an attempt to work out the function of each word.

perceptron

The first artificial neural network, designed in 1957 by Frank Rosenblatt. Artificial neural networks based on the Perceptron have been used in many different applications.

post-synaptic potential

Electric charge induced into the membrane of a neuron when neurotransmitter molecules are released

across a synapse. When the sum of post-synaptic potentials from many synapses reaches a certain threshold, the neuron "fires," producing a stream of pulses along the axon.

program

A set of instructions carried out by a computer's central processing unit (*see* CPU). A program is an expression of an algorithm.

self-organization

The ability of artificial neural networks to find patterns in information without being programmed to look for them.

semantics

The meaning of a sentence or phrase. Semantics are important in our understanding and production of meaningful language, and are a major hurdle in the creation of truly intelligent computers.

strong A.I.

A term used to mean genuine, humanlike intelligence. A computer with Strong A.I. would have a mind and its own independent thoughts.

synapse

The tiny gap between the axon of one biological neuron and the dendrite of another. Nerve signals are carried across a synapse by neurotransmitters.

syntax

The logical structure of sentences, which can be

successfully interpreted and produced by computer programs.

top-down

The approach to artificial intelligence that applies theories about the human mind to computer systems. Top-down systems are successful at tackling problems that are well-defined, such as logic puzzles.

turing test

A scenario involving a computer program holding a conversation with a human being via typed messages. The Turing Test was suggested by mathematician Alan Turing, who believed that any computer that could fool someone into believing it was human could rightfully be called intelligent.

weak A.I.

Intelligence that is not as diverse or flexible as human intelligence, but which shares some of its characteristics. A computer system that has Weak A.I. does not need a mind of its own. Cf. Strong A.I.

weight

A number used in artificial neural networks to represent the strength of the connection between artificial neurons (*see* synapse). To reflect the fact that some inputs to a biological neuron can inhibit the firing of the neuron, some weights are negative.

index

Further reading

Robot: The Future of Flesh and Machines, Rodney A. Brooks, Penguin, 2002.
ISBN 0713995017

Robo Sapiens, Peter Menzel and Faith D'Alusio, MIT Press, 2001.
ISBN 0262632454

The Essence of Artificial Intelligence, Alison Cawsey, Prentice Hall, 1998.
ISBN 0135717795

Virtual Organisms: The Startling World of Artificial Intelligence, Mark Ward, Pan, 2000.
ISBN 0330367102

Darwin Among the Machines, George Dyson, Perseus Books Group, 1999.
ISBN 0140267441

Machine Consciousness, Owen Holland (editor), Imprint Academic, 2002.
ISBN 090784524X

What's the Big Idea? Artificial Intelligence, Jack Challoner, Hodder Children's Books, 1999.
ISBN 0340743824

Online resources

For Cog and Kismet:
http://www.ai.mit.edu/

Chat bots and intelligent agents:
http://www.botspot.com/

Introduction to neural networks:
http://vv.carleton.ca/~neil/neural/neuron.html

20 Questions game with an expert system:
http://www.20q.net/

Loebner Prize:
http://www.loebner.net/Prizef/loebner-prize.html

For Hal:
http://www.a-i.com/

....

Acknowledgments

Illustration
Richard Tibbitts and Martin Woodward, AntBits illustration

Index
Indexing Specialists, Hove

Jacket design
Nathalie Godwin

Picture research
Simon Avery

Picture credits
Aeronautical Systems 6(b); **AMP/Paul Mattock** 29(tr), 59(br); **Apple Computers** 10(b), 30(bl), 33, 54, 55(cr), 64(bl); **Corbis** 28(b); **Department of Cybernetics, University of Reading, UK,** 48-49; **General Atomics Aeronautical Systems, Inc.** 6(b); Genobyte Inc. 52-53; **Hugh Loebner** 8(t); **Kobal** 14 (tr), 22(bl), 53, 60, 63(t), 65(b); **Marco La Civita (Carnegie Mellon University)** 42; **NASA** 5, 50(bl), 62(b); **NASA/JPL/Caltech** 43(b); **Science Photo Library** 8(t), 10(t), 15, 23(br), 25(l), 25(r), 46(t), 50(t), (58), 61(b), 62(c); **Simon Avery & Kree** 9, 12(bl), 13(b), 39(tr), 40(b), 57(br); **Sony Corp.** 2, 8(b); **VoxTec LLC, Annapolis, MD** 30(tl);

Game of Life applet (p. 47) by Edwin Martin: www.storm.org/gameoflife

Every effort has been made to trace the copyright holders.
The publisher apologizes for any unintentional omissions and would be pleased, in such cases, to place an acknowledgment in future editions of this book.

All other images © Dorling Kindersley.
For further information see: **www.dkimages.com**